SENSATIONS ABOUT FLAT EARTH

by LUCA BERTORELLI

Title | Sensations about flat Earth

Author | Luca Bertorelli

ISBN | 978-88-31685-18-4

Youcanprint Via Marco Biagi 6, 73100 Lecce www.youcanprint.it info@youcanprint.it

CHAPTER 1

Pac-man vs Asteroids

Speaking positively or negatively about a subject conveys the basic information about it anyway. In a television report, the "Pac-man effect" was attributed to Flat Earthers, in other words the explanation of what would happen to an airplane if it crossed the boundary of a hypothetical flat Earth: if it were thrown off the edge of the Earth, it would return from the opposite side. First, I went to see if there were supporters of the flat Earth in the world. The English inventor Samuel Birley Rowbotham in 1849, under the pseudonym "Parallax", published a sixteen-page booklet entitled Zetetic Astronomy: Earth Not a Globe. In 1956 Samuel Shenton, a member of the Royal Astronomical Society, founded the Flat Earth Society. So not only did the Flat Earthers exist, they also had their own association. I went on to examine "the Pac-man effect" without finding a single supporter of the Flat Earth who would claim such a ridiculous thing. To my amazement, I discovered that it was instead claimed by an official science scientist in the documentary series presented by Morgan Freeman.

In the episode entitled "Do the boundaries of the Universe exist?" (episode 2, season 2), the astronomer Jean-Pierre Luminet hypothesized a finite Universe delimited by a boundary. He described it as an asymmetrical dodecahedron, with rounded faces, more like a soccer ball. He argued that if you came out of the Universe from one of these faces, you would return to the opposite side, but rotated three-hundred and sixty degrees. Jean-Pierre Luminet was inspired by Asteroids, a famous video game that was fashionable in the 1980s, made by Atari in 1979.

I continued my research and came across an article that exposed the belief in the flat Earth. Although it was published by Repubblica, a respected newspaper, I expected to read it with the expectation that I was facing a colossal hoax. The author listed concepts that were completely foreign to me and allowed me to discover a new way of thinking aimed at redefining the shape of our planet. At the end of the article I found myself reflecting on how the awareness of not living on a globe would affect my life. It probably would not change my days divided between home, work and the journey that unites them; I would still do the shopping and make the most of my free time. However, I began to feel a sentiment of uneasiness: are we living at the dawn of the third millennium and still debating the true shape of our planet? The fact that the Earth had always been considered flat by our ancestors is no mystery but proposing it again in today's world seemed like a strange idea.

The thing is: either the Earth's shape is spherical or it's flat, easy, isn't it? But a problem arises because while in the first case it would be all already written and consolidated, in the case of the latter, the foundations of science and parts of our history would be undermined. Moreover, if it were true, someone would not only have been lying to you and me for decades, but they would have done it in such a shameless way that the whole of humanity would now suffer for their lie. From my point of view, the reason for such devious deception seemed mysterious and impenetrable. However, I decided to devote a little more time to the theories of the Flat Earthers and I spent many hours, in the days that followed, unravelling thousands of articles and videos, both for and against the flat Earth. Much of the material was more inclined to ridicule the Flat Earthers than providing concrete scientific explanations. The only certainty that everyone agreed

SAYS EARTH IS ROUND

AND HE MAY BE THROWN INTO PRISON.

Sad Condition of Affairs In England— Sir John Gorst Accused of Intention to Teach False Precepts—City of Portsmouth Excited.

It is painful to read that Sir John Gorst, the head of the British educational department, is in serious trouble and has been threatened by Mr. Ebenezer Breach and other taxpayers of the city of Portsmouth, in the kingdom of England, with prosecution under the "imposters' act." It seems that the schools of Portsmouth have been teaching the damnable and heretical doctrine that the earth is a sphere. Sir John's attention has been called to this dissemination of seditious and treasonable doctrine, but he has refused to correct the abuse. Ebenezer and his friends know, of course, that the earth is as flat as a pancake. They have been patient with Sir John, and day after day have allowed the false teaching regarding the shape of the earth to go on, but can stand it no longer, they say, to see their children corrupted with this most "heretical doctrine," as the complainants call it in this protest. Sir John Gorst has many on was the presence of an atmosphere or dome that separates us from outer space. In some ancient texts it has been claimed that the "As above, so below" and that beyond the dome there is more water. But if the Earth is flat what is below? Even this legitimate question was not answered except for other ancient texts where it was claimed that "the tree of life", with deep roots, reaches the world below.

I was almost convinced to let it go, when I read an article dated April 1900 in which a professor from Portsmouth, England risked going to prison on charges of teaching false precepts in the school system. In short, this Sir John Gorst, head of the British education department, had been sued for "an act of imposture" because "it seemed that in the school in Portsmouth they taught the damn heretical doctrine that the Earth is a sphere" (The Cook County Herald, Grand Marais, Minnesota, Saturday April 21, 1900). It was a turning point for me, the confirmation that something was not right at all.

Think of this quote: "everything that is taught in the school system during a specific generation becomes the truth, whether it is true or not. This means, if the flat Earth was taught in schools at a time relatively close to ours, that State was preparing that generation for a future event of enormous magnitude. The first world conflict would begin only a few years later, at the very expense of that generation, confirming the preventive choice adopted by the State in question. Imagine if during the conflict the equidistant azimuthal map had proved to be wrong or incomplete. The planes would not have accurately bombed places considered strategic, or parachuted soldiers into predetermined areas, inside or outside enemy lines or even navigate navy forces in safe waters. According to the U.S.G.S. (United States Geological Survey) the equidistant azimuth maps are detailed and accurate.

Samuel Birley Rowbotham

CHAPTER 2
The Map of Gleason

With internet's help I found an azimuth map and downloaded it into my PC, printed it out and, once it was laminated, I examined it carefully. In the middle of the map there is the North Pole with all the continents arranged around it, while the circumference is formed by the South Pole, which sets the limit all around it. The two tropics (Cancer and Capricorn), with the Equator between them, stand out on the lands and the seas above them in three concentric circles. All the war plans, belonging to the First and Second World War, have been prepared by referring to an equidistant azimuthal map. The most accredited map is that of Alexander Gleason, a cartographer who published it in 1892. The map is deposited and preserved in the Boston Public Library and was created by the Buffalo Electrotype and Engraving Co. Immediately after the end of World War II the U.N.O. had been created, on the 24th of October 1945. In the symbol adopted by the United Nations a flat map of the Earth stands out in the

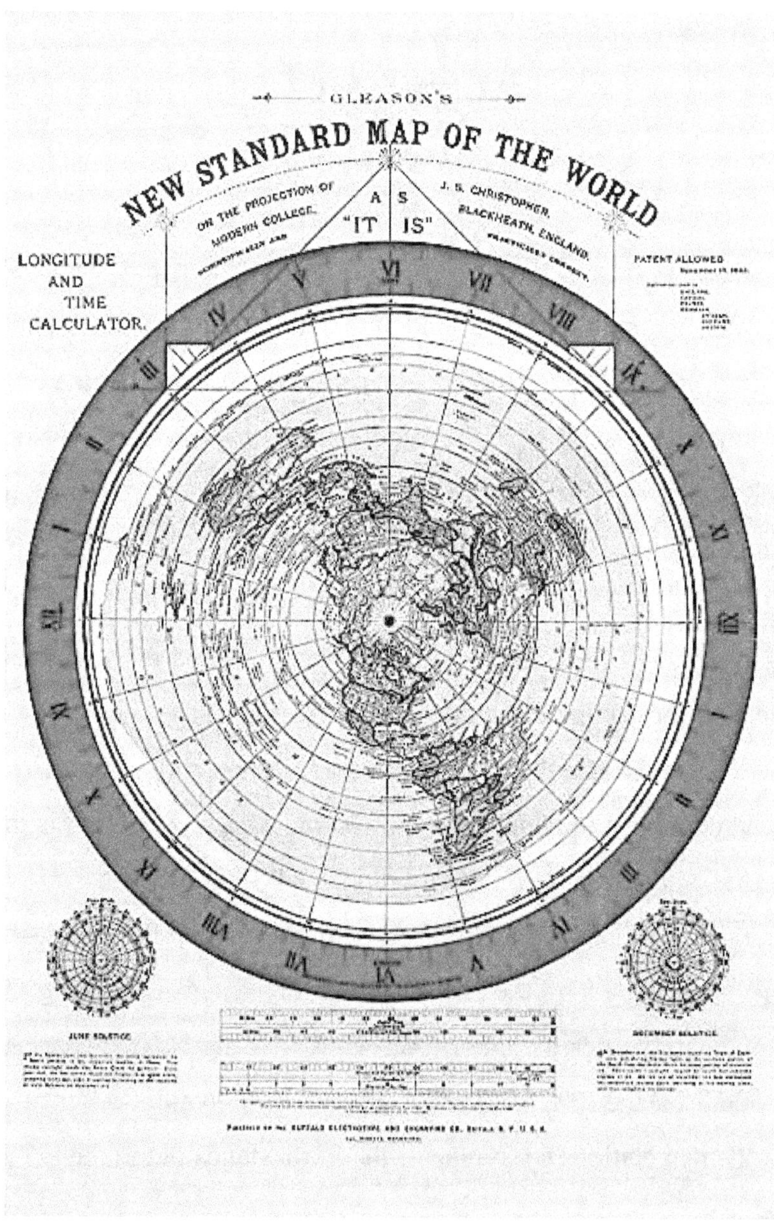
GLEASON'S
NEW STANDARD MAP OF THE WORLD
ON THE PROJECTION OF
J. G. CHRISTOPHER,
"IT IS"
BLACKHEATH, ENGLAND.
LONGITUDE
AND
TIME
CALCULATOR.
PATENT ALLOWED
JUNE SOLSTICE
DECEMBER SOLSTICE

centre, perhaps the map of Gleason, since it matches with it. Sometimes the best way to hide something is to put it in plain sight.

The fact that even in recent times the Earth was not considered a sphere led me to new scenarios. Perhaps not all the teachings people inculcated us when we were children can be considered absolute truths. We all have a duty not to trust anyone without first verifying and trying to think individually without following the most common thinking of the masses, especially those led by people who govern them. When I talk about the masses, I also mean you and me. Sylvain Timsit describes very well how easy it is to exercise control over the masses. For example, one way would consist of diverting attention from serious problems to lighter subjects. I belong to this case study because I am a sports fan and certainly, if I had to choose between a political programme and a sport programme, I would opt for the latter. The fact that we communicate less with each other is another element against us. It is enough to look around us in a crowded area to understand that everyone is distracted, isolated from the others, focused on their smartphone, iPhone, iPad or tablet. If you are a bit like me, don't despair; being aware of this partly diminishes the effect of this mental control on us. You may be wondering the reason of my apparent digression, but I can assure you it is pertinent: as a matter of fact, I have not been able to find a single debate between Flat Earthers and heliocentric people.

I have found out that only supporters of the flat Earth have repeatedly asked for an open confrontation with the scientists of official science, but they have always been ignored. Ignored all over the planet, I mean. This happens for two reasons: first, inviting them into a broadcast for a debate would give them visibility; second, it is easier to ignore a problem than to solve it.

It is of great comfort to me to know that many scholars and scientists, great minds of the past have developed during the past centuries the same questions as today and have tried to give an answer to them. If you also feel the urge to go deeper into some topic and do not want to run into videos broadcasted by Google, you will find in the credits all the books and sites I have researched. You can have an easy feedback of both the characters and the facts described in this book. If you do not have any time due to the frenzy that characterizes our era, do not despair, I have done the work for you!

William Carpenter
(1830-1896)

One of the first modern supporters of
the flat earth.

CHAPTER 3

The balance of number 7

Every religion preserves the knowledge of ancient sacred texts, it would be impossible for me to describe the content of the millenary knowledge contained in them and condense it into a chapter. Since this book is neither a text of history nor of theology, you will understand the why I made the choice to quote only a part of some ancient texts. The most widespread and translated book in the world is the Bible, handed down first verbally and then written. Although the Bible was translated several times over the ages into several languages, some of them now extinct, it remains almost unchanged both in its syntax and in its concepts. The first Bible was printed in the German city of Mainz by John Gutenberg in 1455, the first person who used metallic movable characters. The Jewish printers in Soncino (Italy) printed the first complete Hebrew Bible in 1488. The explanation that the Bible offers to the origin of life has been handed down since the dawn of time to the present day, lasting over millennia. If you have never read this passage, you check Genesis, we only need to make a list of the days of the Creation:

First day - Night and Day

Second day - Sky and Sea

Third day - Trees and plants

Fourth day - Sun and Moon

Fifth day - Fish and birds

Sixth day - Man and animals

Seventh day - Rest

Seven days, like the seven wandering stars classified for the first time in ancient times we know today as:

Sun-Sunday

Moon-Monday

Mars-Tuesday, in italian: Marte, Martedì

Mercury-Wednesday, in italian: Mercurio, Mercoledì

Jupiter-Thursday, in italian: Giove, Giovedì

Venus-Friday, in italian Venere, Venerdì

Saturn-Saturday

In the ancient world, these were considered seven planets while Earth, the only suitable place for life, was not considered one of them. According to science, it all began with the "Big Bang", a colossal deflagration from nowhere, which is still expanding.

Today it is considered valid as only a theory. As far as the origin of life in the Universe is concerned, some scientists hypothesize the phenomenon of panspermia, according to which life, wherever it was formed, having a common denominator that acts as a vector, would originate on several planets in different galaxies. In fact, it is a theory that takes away our Earth's unique role in the Universe. According to Drake's equation, it has been established that it is mathematically possible to find other Earth-like planets in the Universe.

Also, in the Genesis it is mentioned Enoch, father of Methuselah, patriarch kidnapped by God (Genesis chapter 5, verses 21 to 24).

Fragments of his book were found in 1947 in the Dead Sea Scrolls. Enoch, in the first book (33 from 2 to 4), talks about the fact that he saw the end of the Earth on which the sky rests and, beyond it, a horrible and dark place without neither sky, nor water, nor birds, where the seven wandering stars are paying for God's punishment because they disobeyed Him by not arriving at the precise time He established. In Enoch's book the seven stars are identified in the angels of the Seven Churches. In the astronomy of the ancient Greeks we find the seven wandering planets in this order: Moon, Mercury, Venus, Sun, Mars, Jupiter, Saturn.

In the same order the ancients Egyptians catalogued them as seven planets depicted on seven concentric circles with the Earth at the centre. In the Book of Creation of Sefer Yetzirah, considered the oldest cabalistic text in the world, the seven planets are described in reverse order: Saturn, Jupiter, Mars, Sun, Venus, Mercury, Moon.

The number seven appears countless times in many texts, from the Bible to book 777 by Aleister Crowley: there are seven days in the year, seven days in the week, seven gates of the soul, male and female (2 eyes, 2 ears, 2 nostrils, 1 mouth), seven chakras, seven deadly sins (anger, greed, envy, pride, gluttony, sloth, lust), seven universes; seven is the Buddhist number of completeness, seven lands, seven seas, seven rivers, seven deserts, seven weeks, seven years, seven sabbaticals, seven musical notes, seven jubilees, seven beloved in Paradise, seven seals, seven alchemical metals, seven firmaments; seven are the gifts of the Holy Spirit in Christianity: wisdom, intellect, counsel, fortitude, science, piety and the fear of God. .. There are many others, but I conclude with the Seven Wonders of the World which, according to Manly P. Hall, were not found by chance. In fact, he argues that they are symbolic structures and for this reason placed in specific positions, built by the widows of the seven sons in connection with the seven planets in honour of the seven planetary genes.

Only a Freemason initiate can understand the identical secret symbolism of the seven seals of revelation and the Seven Churches of Asia. For those who would like to learn more, read the book: The Secret Teachings of All Ages, secret teachings for all ages, by Manly P. Hall, a thirty-third degree Freemason. We have seen together the huge importance the number seven has for humanity: it is considered the number of philosophy and analysis, but also of loneliness and completeness.

CHAPTER 4

From Flat Earth to Sphere

Religion and science have always been at odds with each other trying to give meaning to fundamental concepts such as life, its meaning, death and even what we can expect after death. One can accept a religious concept by faith but certainly not a scientific one, which requires solid and proven foundations.

Eratosthenes (about 276 B.C. - 194 B.C.) was the first to calculate the circumference of the Earth as a sphere. Experts agreed on the procedure adopted but doubted the actual accuracy of its implementation. Eratosthenes had the objective of measuring the difference in the inclination of the shadow between two points that were distant from each other 5000 stages (about 785 km), a real challenge for the time. His choice fell on two cities: Alexandria and Syene (the current Aswan). The unit of measurement expressed in stadiums, where one stadium would correspond to 177.6 metres, appears to experts to be inaccurate for measuring long distances. Another doubt is given by the absence of reliable watches that can carry out measurements synchronously. At the time, sundials were used, therefore the precision required to ensure the success of the experiment was further reduced.

Some people argue that it took two people to do both surveys at the same time, but this may not have been the case. Eratosthenes was aware of a well in Syene where the Sun, at a certain time of day, on a certain day of the year, was exactly perpendicular to it, at zero degrees of inclination. With this certain datum in his hand, alone, it would have been enough for him to be in Alexandria at the same time of the same day to make the second measurement. Despite what really happened, Eratosthenes quantified the variation of the shadow between the two points in 7 degrees and 12 minutes. He then calculated the circumference of the Earth by applying Euclid's geometry. I go on to deal with this subject later, but I must anticipate that, because of its considerable distance from the Earth, the Sun's rays should be parallel. Surely it has also happened to you, you see the rays of the Sun filtering through a cloudy sky, well, the rays diverge so the source of light appears much closer than stated in science books. For Flat Earthers it is just like this, the Sun is closer.

According to our scientists, however, the Earth's atmosphere refracts sunlight and making its rays diverge. This last explanation, however, is not convincing because the atmosphere is convex and, according to the law of physics, a lens always makes the rays of light converge, not diverge. Here is where we have the problem: either the atmosphere always refracts the Sun's rays making them diverge and Eratosthenes' calculations were wrong, or the atmosphere does not refract its rays and Eratosthenes' calculations are correct, but in this case the Sun should be much closer than stated in our science books.

Eratostene

CHAPTER 5

The Earth in the Solar System

Part 1

In a more poetic vision of the origin of the Universe, it is thought that the original North Star exploded giving rise to the present North Star, the firmament and all of us: we are considered stars, each of us with our own "self". This recalls a little bit the Buddhist philosophy, Microcosm in the Macrocosm. In the Bible, God counts and numbers the stars giving them a name, God also counts the grains of sand but does not name them because only the stars are considered alive.

From the Big Bang to today about 13 and a half billion years have passed. Some scientists claim, with the help of scientific satellites (Cobe, Wmap, Plank), that they can still see the "cosmic background radiation", a remnant of the early stages of the birth of the Universe. When you explore Space, it is like looking into the past and the more the instrument is progressive the more you can go backwards in time. From 1978 till today, four Nobel prizes have been awarded for this discovery, so to say the oldest photons we can ever hope to see, dating back about 380,000 years after the Big Bang.

Six thousand years before Christ the Sumerians described a Solar System formed by the Sun in the middle, 10 planets and the Moon. The tenth planet, called Nibiru, was visible every 3600 years because of its long orbit and it was thought to be placed between Mars and Jupiter. The Sumerians claimed that the Annunaki, inhabitants of Nibiru, had contributed to the evolution of homo erectus into homo sapiens. In antiquity, Sumerians aside, it was believed that the Earth was the centre of everything, and it was taken as read to explain the Solar System by combining Aristotle's cosmological system with Ptolemy's astronomical system. Nicholas Copernicus in 1543, shortly before his death, published the book "The revolutions of the celestial bodies" elaborating the theory postulated by Aristarchus of Samos (3rd century B.C.) and then consolidated over time as the most accredited heliocentric theory.

(Really astonishing!)

Let's start by saying that the Earth rotates at about 1700 km per hour, while the speed of sound is 1191 km per hour, or Mach 1. This fact may concern you, but it is nothing compared to the speed at which the Earth orbits the Sun: 107000 km per hour, almost Mach 90. Finally, our Solar System orbits our Galaxy at 772500 km/h, Mach 640.

Let's take the place you are right now for example: if the rotation of the Earth corresponds to the direction of its orbit you could reach 108700 km/h. If you went to the opposite direction, over the next 12 hours, you would decelerate to 105300 km/h. The angular and linear moment should also be considered, including variations due to the Earth's elliptical orbit in its annual path. Without doing it the hard way with too many difficult calculations, let's just bear in mind that the difference between the two speeds reached is enormous: 3400 km/h, almost three times the speed of sound. All of us who inhabit the surface of the Earth should perceive and undergo these continuous as accelerations and decelerations at regular intervals of twelve hours. Now let's imagine that you are at the Equator at the point of the greatest moment at +1700 km per hour and that on the opposite side, at the same time, you can find me at 1700 km per hour. We know that gravity is the same all over the planet, but in this case, you would be projected towards space while I, being at the antipodes, would struggle to move. We must then assume that gravity has an intelligence of its own and that it is able to modulate itself according to our position on Earth.

Part 2

There is an element of total disagreement between heliocentrism supporters and Flat Earthers that seems to have no solution: the sidereal day. Each side claims that it is pointless to argue with the other side. I find this diatribe amusing and I will try to explain both points of view. If you have pen and paper you can do it with me by drawing a circle in the middle of a piece of paper, our Sun, then a circle at the bottom, at 6 o'clock, representing the Earth "A". Let's darken the half of the Earth "A", which is in shadow. The Earth orbiting the Sun after about 180 days will be on the opposite side, at 12 o'clock. We draw a circle at that point representing the Earth "B" and darken the half in shadow again. Now we have Earth "A" with the shadow zone at the bottom and Earth "B" with the shadow zone at the top and the Sun in the middle. Let's suppose that where we are right now is noon, so let's put a cross on Earth "A" in the circumference part illuminated by the Sun, right in the middle. Now we put a cross on Earth "B" in the same place, but you can see that it is in the shadow at midnight. To Flat Earthers the Earth is stationary for this very reason. Official science provides an explanation to analyse as to why even after 180 days, it is still noon for us.

Shortly, they tell us that the Earth's rotation does not take 24 hours, but 23 hours and 56 minutes. The missing 4 minutes, which completes the 24 hours of a day, are given by the revolution motion of the Earth around the Sun. Taking as reference the example from before, based on a time span of 180 days, if we multiply those 4 minutes by 180 we obtain 720 minutes. If we then divide them by 60, we get the missing 12 hours. Here we are after 180 days on Earth "B" at noon, as on Earth "A", but upside down, so cross is below! This is a "trick" by heliocentrism supporters, in the Flat Earther's opinion. In fact, by subtracting time from the Earth's rotation to add it to the motion, they square their basic thought with mathematics, that is the Earth is in motion. Even the stars contradict heliocentrism because, six months later, we should be looking at a totally different starry landscape.

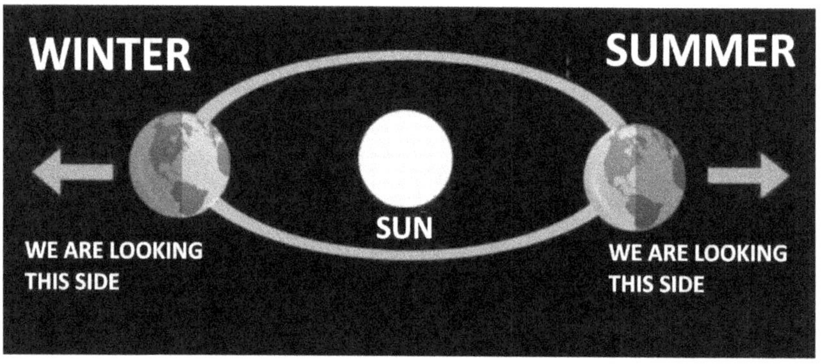

CHAPTER 6
Curvature and horizon

No matter how life originates and develops in the Universe, light, same as water, is considered a fundamental ingredient. It is assumed that it came to Earth from space through vectors and that it is one and a half billion years old. The human body is composed of about 60% water, while more than 70% of the surface of our planet is covered by it. With its two hydrogen molecules bound to one oxygen molecule, water is very versatile. At high temperatures the molecules are divided, and the water evaporates, releasing power, while at low temperatures it is solidified in ice. Another property of water is that it always takes the shape of the object that contains it while remaining level on the surface. No matter the size of the container, be it a glass, a tank, a lake, the water will take the shape and level itself on the surface. One would think that this extraordinary property of water also applies to the sea. But this does not happen: our science argues that the sea, held back by the force of gravity, bends all around the Earth.

From the "Earth Curve Calculator" site you can calculate the curvature from the distance separating two spots. There is also a practical application for Android, Arflat TCC, in order to do that. Its operation is identical to the Earth Curve, plus it provides optical refraction data.

Trip on the tower

On Sunday, November 18th, 2018, at 6 p.m., the Panoramic Tower was inaugurated in Pesaro announcing: "on days of good visibility this tower will allow you to see Rimini, the coasts of Croatia and Mount Conero". The tower, built for the 150th anniversary of the death of Gioachino Rossini, was dismantled on September 16th, 2019. Now let's take the Arflat application or Earth Curve Calculator to calculate the curvature of the Earth. First, we set the height of the tower, 75 meters, then the minimum distance between Pesaro and the Croatian coast, that is 123 km.

Croatia's coastline should be 665 meters below the horizon line, 540 meters with refraction, 815 meters applying reverse refraction. Thousands of tourists have visited the Panorama Tower and they observed the coast of Croatia without any problems. Many measurements were taken all over the planet, even with infrared lenses, but no curvature was detected. From the keel the boats seem to be swallowed up by the horizon, but with the help of powerful lenses they always become visible again. This is also achieved by improving the viewing angle by climbing higher. This phenomenon is due to an optical effect given by the perspective, and the conclusions of Focus Junior, in the section "Flat Earth: 5 elementary tests that break the hoax down!", are to be considered completely wrong. You cannot see behind the curve below the horizon, that means the boat is still there and the sea does not curve at all. More evidence that the sea does not curve at all is provided by the submarines. It would be impossible for them to stay for hours at periscope depth with the rudder at zero degrees, about 9 metres below sea level, as they would have to periodically compensate for the curvature of the water in order not to re-emerge. This precaution seems not to be necessary for underwater navigation and is also a valid concept for aeroplane pilots, who would have to constantly correct their altitude by turning downwards in order not to fly straight into 'outer space'.

Let's go to court

Mr. Zen Garcia offered a \$15,000 prize and challenged anyone to provide proof of the Earth's curvature as well as proof that the Earth rotates above 50 mph, about 80 kilometres per hour. Mr. William Menke Thomson, after having done the calculations, demanded that money from Zen Garcia. The latter, unconvinced, did not pay him and forced Mr. Thomson to file a lawsuit against him. The notification reached both parties on May 16th, 2019. By judgement of June 11th, 2019, the Barrow County Court, in the state of Georgia, ruled in favour of Zen Garcia. Does this mean no curvature? No rotation? Perhaps Mr Thomson, no doubt attracted by the money prize, was not the most qualified person to fulfil the task. But then why was he not supported in court by any professor or scientist for his thesis?

An article in the Seattle Star, published on August 17th, 1921, describes the theories of Flat Earth by Charles Bishop, an amateur scientist. The elderly gentleman went to Woodland Park and, after showing a map of the Flat Earth he had created, he exposed his theories to visitors, challenging any colleague to prove him wrong. He had studied and spent twenty-five years developing these theories.

Speaking about Magellan, when he was asked how a ship could circumnavigate the continents on a flat Earth, he answered by bending a twig and placing it on the map: - That's how!

I am now mentioning this article because Bishop states here a concept that struck me a lot when I read it. He claimed with certainty that on a flat Earth there must be more land than water, practically a basin of greater size to contain the oceans. This concept would imply the existence of huge expanses of unexplored land beyond the circumference, i.e. beyond Antarctica. This statement by Charles Bishop gives credence to the statements, following the expedition to Antarctica in 1929, of Admiral Byrd, a well-known explorer of the two poles I will introduce later. Another mistake Bishop attributed to us was the fact that we consider the Earth in motion: he did not believe that the force of gravity could hold back water. He claimed that if gravity was strong enough to hold the oceans, to compensate for the centrifugal force due to the Earth's rotation, we all wouldn't be able to move. He then deduced that the Sun moves at 1000 miles per hour and that it is not a ball of fire but of electricity. Bishop also claimed that he could prove that the Sun is only 3000 miles above the Earth with a radius of 2 miles (about 5000 km from the Earth with a diameter of 3.2 km). The scientist Hiromichi Ilda, his Japanese contemporary, came to similar conclusions. He believed he was ready to demonstrate at the Peace Exhibition in Tokyo that "the Earth is a flat extension of land and water and that there are still unknown lands beyond the Antarctic Circle".

CHAPTER 7
Rotation I

A year after his death, Galileo Galilei regretted the heliocentric choice he had made even before the Pope. These were dark years for Copernicus' theories where many men lost their lives to defend them. An example of this is the philosopher Giordano Bruno who was burned at the stake in February 1600. In the nineteenth century heliocentrism still made its way, but the motion of the Earth was still to be verified. At the end of the nineteenth century physicists based their theories on the assumption that space was composed of ether, a substance through which light and other electromagnetic waves could propagate. All attempts to measure the speed of the Earth's movement around the Sun through the ether resulted in zero. In 1887 the two physicists Michelson and Morley prepared an experiment that was considered by many physicists the most important experiment in physics. The idea was that the Earth, moving in space through the ether, would create an "ethereal wind" on its surface. By projecting a light into this ethereal wind, this light was expected to move slower than a light projected through it.

For example: if you swim in a river, you move faster if you follow the current instead of swimming in the opposite direction.

The Michelson-Morley experiment did not prove that the Earth was moving, only that the light was not affected by the ethereal wind. Albert Michelson declared: - These conclusions directly contradict the explanation that assumes that the Earth is moving -. Science found itself at a turning point: either discard the ether idea or admit that the Earth does not orbit the Sun. Einstein completely discarded the concept of ether from a physicists point

of view and in 1905 he wrote an article about "narrow relativity",
also known as special relativity. The principles of narrow
relativity are:

(principle of relativity)

1: all physical laws are the same in all inertial reference
systems.

(constancy of the speed of light)

2: The speed of light in vacuum has the same value in all
inertial reference systems, regardless of the speed of the observer
or source.

The second principle satisfies the "Lorentz transformations"
named after him, even though Joseph Larmor discovered them in
1897. They are transformations of coordinates that make it
possible to describe how the measure of time and space between
two rectilinear points varies. In 1913, George Sagnac assembled a
source of light on a revolving table, which passed through a
splitter and split into two rays. Through mirrors, these two rays
bounced in opposite directions until they returned to the splitter
and recombined. Once recombined, they proceeded to be
immortalized on a photographic plate. At a stationary table no

interference was detected, but by rotating the table 2 turns per second, the rays of light generated interference fringes on the photographic plate. This meant that with the table in motion an asymmetry was created because one ray had to travel a longer distance than the other. Sagnac's experiment showed that light can vary in speed, despite Einstein's second postulate of relativity.

But there is something more: Einstein had ignored the ether because it was inconvenient to his theory, Sagnac proved that the ether existed, existed and had an influence on light.

This experiment will be ignored (who knows why!) by Einstein, who will present the theory of general relativity two years later and will present it in great fashion in the form of lectures at the Prussian Academy of Sciences in Berlin. The lectures began on November 25th, 1915.

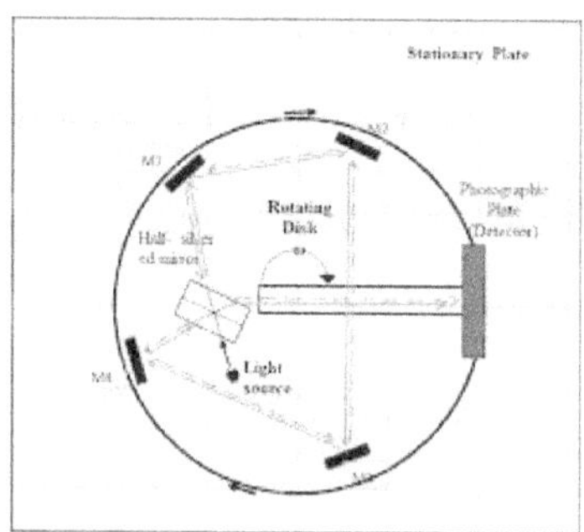

CHAPTER 8
Rotation II

Any object that leaves the earth's soil is subject to the force of Coriolis, which takes its name from the French physicist who first described it in 1835. It is believed that the effect of Coriolis is at the basis of the formation of cyclonic or anticyclonic systems in the atmosphere. Coriolis had the intuition to study the effect of an object leaving a surface in motion. If we were jumping vertically on a carousel: we would end up falling back on a different point from the starting one. We are too small compared to the Earth to see the effects of this, but with a bullet or a plane it should be possible. During a conflict, long-range guns may miss their target if they are several kilometres away. An odd fact is that in military ballistics the aiming corrections due to Coriolis' force are not considered. The same principle applies to planes once they take off. I'm wondering how they would land on a terrain that moves at about 1,000 miles an hour. Even assuming the plane had the rotation of the Earth in favour, at what speed should the plane land?

Fifteen N.A.S.A. documents dated 1988 have been recently appeared: in these documents the results of calculations for the flight of aircraft and helicopters are reported. In all the documents this sentence appears; - Aircraft flying in a stationary atmosphere flying on a stationary, non-rotating surface -. I am now quoting the user manual for linear aircraft model, last page: - The non-linear motion equations used are equations of 6 degrees of freedom, stationary atmosphere and flat, non-rotating ground hypothesis -.

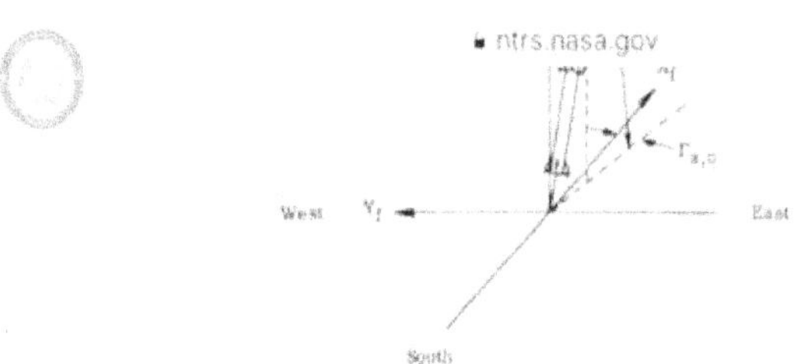

Figure 1.- Earth-fixed axis system and body axis system at vehicle lift-off.

Transformation 2

The radar provides range, azimuth, and elevation data from the radar site to the vehicle throughout the flight. These data in conjunction with wind data can be easily converted to the pitch γ_p and yaw γ_y flight-path angles. (See appendix.) It is assumed herein that the earth is represented as a flat, nonrotating reference frame. (This assumption limits the analysis to short-range, short-duration flights.) The free-stream

NASA
National Aeronautics and
Space Administration

Report Documentation Page

1. Report No.	2. Government Accession No.	3. Recipient's Catalog No.
NASA RP-1207		

4. Title and Subtitle		5. Report Date
Derivation and Definition of a Linear Aircraft Model		August 1988
		6. Performing Organization Code

7. Author(s)	8. Performing Organization Report No.
Eugene L. Duke, Robert F. Antoniewicz, and Keith D. Krambeer	H-1391
	10. Work Unit No.
9. Performing Organization Name and Address	RTOP 505-66-11
NASA Ames Research Center Dryden Flight Research Facility P.O. Box 273, Edwards, CA 93523-5000	11. Contract or Grant No.
12. Sponsoring Agency Name and Address	13. Type of Report and Period Covered
	Reference Publication
National Aeronautics and Space Administration Washington, DC 20546	14. Sponsoring Agency Code

15. Supplementary Notes

16. Abstract

This report documents the derivation and definition of a linear aircraft model for a rigid aircraft of constant mass flying over a flat, nonrotating earth. The derivation makes no assumptions of reference trajectory or vehicle symmetry. The linear system equations are derived and evaluated along a general trajectory and include both aircraft dynamics and observation variables.

17. Key Words (Suggested by Author(s))	18. Distribution Statement
Aircraft models Flight controls Flight dynamics Linear models	Unclassified — Unlimited Subject category 08

19. Security Classif. (of this report)	20. Security Classif. (of this page)	21. No. of pages	22. Price
Unclassified	Unclassified	108	A06

NASA FORM 1626 OCT 86

*For sale by the National Technical Information Service, Springfield, VA 22161-2171.

NASA-Langley, 1988

Let's go back to the subject: the aberration of light studied by the English astronomer James Bradley gives us a good starting point to continue the main topic. Think about when it rains above us: the water drops perpendicular to us and the ground; but if we move, the drops seem to change angle and they move towards us. If we point a telescope at a star, and both the telescope and the star are motionless, the light comes right into the telescope. If one of them is moving, you must tilt the telescope the necessary angle in order to compensate for its motion. But how can we tell if the telescope on Earth is moving and not the star? In 1871, Sir Biddle Airy performed an experiment that Bradley had already devised in 1729, but never put it into operation. Airy noted the inclination needed to make a star's light fall into the centre of the telescope, so he filled it with water and returned to observe the star. The water slowed down the light, so if the telescope was moving, it would have to be tilted more to achieve alignment with the star again. Airy found the same angle with or without water in the telescope. He expected a difference of 30 seconds of arc but detected only 0.8 seconds. It's all in his report to the Royal Society of London dated November 17th, 1871; a four-page report that shows that the Earth is stationary.

This experiment, also known as "Airy's failure", is not taught in schools and was also removed by Wikipedia, which described it wrongly, by attributing to Airy the only merit of having tried to prove the variation of stellar aberration.

Sir Biddle Airy

We are often inclined to believe that people who lived long before the modern era were less qualified than they are today in their respective roles. Sir George Biddle Airy had a mathematical mind and was not accustomed to technical problems, yet he was the first to design corrective lenses for astigmatism, a defect he himself suffered from. In 1828 he applied for the prestigious Plumiana Chair of Astronomy with a salary of £500, versus the

£300 of his predecessor, and he managed to achieve both aims. He was head of the Committee on Weights and Measures in 1834, he was elected to the Royal Society of Edinburgh in 1835, to the Royal Society of London in 1836, then he was President of the Royal Astronomical Society in 1845, of the British Association in 1851 and of the Royal Society from 1871 to 1873; during this period he was the first to discover stellar aberration. He was awarded many other honours, including the gold medal of the Royal Astronomical Society in 1833 and 1846.

CHAPTER 9
Gravity

Isaac Newton is considered one of the greatest scientists of all time; he was President of the Royal Society. In 1687 he published "Philosophiae Naturalis Principia Mathematica" describing the law of universal gravitation. Before the end of 1800 many scientists had detected the error of Newton's gravitational theory applied to the Universe because this theory was not able to explain the movement of Mercury around the Sun. It was Albert Einstein in 1915 who formulated the theory of general relativity sure to give a solution to Newton's omissions. Einstein, in practice, argued that gravity is much more than a force: it is a curvature of the space-time continuum. With Einstein's "tensor" (a mathematical formula) he expressed the curvature of space-time in the field equation for gravitation in the theory of general relativity.

Between density and mass

No one has offered proof that it is the mass of the Earth that attracts objects to itself; when an object falls this is called acceleration. The denser things fall while the less dense float. The British scientist Henry Cavendish was the first to create a laboratory experiment to measure the force of gravity between masses.

Henry Cavendish

Cavendish's experiment, carried out in the years 1797-1798, consisted of a torsion balance instrument consisting of two lead balls connected to a beam hanging from a wire. By bringing the torsion balance instrument closer to other masses, Cavendish detected the oscillation caused by each pair and modified the formula of that time by substituting "G" of gravity into "m" of mass.

But the real value of "m" is unknown:

It is a theoretical value that defines something intangible. Einstein proposed the value of "m" in his most famous formula where "E" equals "m" multiplied by the square of "c". Theoretical values within the theories. Nikola Tesla claimed that it is magnetic radiation that attracts objects, which fall towards a greater energy. Not sharing Einstein's thought of relativity, Tesla declared: - I maintain that Space cannot bend for the simple reason that it cannot have properties -. A thorn in Einstein's side and in the scientists, who nowadays justify their existence by talking about "dark matter". Be careful of your choice of words: "matter" has properties, but it is "dark" and therefore impossible to prove its existence. I wonder if today, as then, they would let Nikola Tesla die alone, marginalized by everyone and in poverty.

Antigravity

A friend of Tesla, Edward Leedskalnin, built a coral castle made of huge blocks, from 1923 to 1951 some weighing over twenty tons. He built it with a simple wooden tripod which, according to experts, cannot lift such weights.

Edward Leedskalnin

Moreover, the man was only one meter and thirty-nine tall, and yet, after building the castle, not being happy with its location, he dismantled it and then rebuilt it a few miles away: Coral Castle, Homestead, Florida. Edward claimed that he knew how the Egyptians and Mayans had built the pyramids. He said he was ready to tell his friends this secret, but he died shortly before he confided to them. From the photos of the time you can see a mysterious box on the top of the tripod, probably containing a device that could cancel the weight of the coral. The inhabitants nearby Coral Castle claimed that they used to hear a disturbing noise, which used to accompany Ed's nightly hours of work. The same gloomy noise heard by Tesla's neighbours before he created an earthquake in New York City in 1893. Tesla's Oscillator is a generator of alternating motion electricity. Perhaps Tesla had helped his friend Ed in his quest by providing him with the inspiration to create an instrument like his own, capable of moving huge blocks of coral.

That would explain why Edward chose to work alone and at night. One thing is certain: the two friends had specific and avant-garde knowledge, not only compared to the era when they lived, but also compared to our time. Just think of the Telefunken Wireless built by Tesla, the first antenna of its kind, dismantled by the military on July 4th, 1917, during the First World War, or his studies on "vacuum tubes", which generated X-rays. The electrical engineer Eric Dollard, who will be introduced in Chapter 11, was fascinated by both Tesla and Ed Leedskalnin. He developed his own theory about Coral Castle: "Coral Castle Quadrant Theory".

Our science takes it for granted that, thanks to the force of gravity, the planets of our Solar System have orbited for millions and millions of years in precise orbits. This contradicts the second law of thermodynamics, which implicitly goes against perpetual motion. Edward Leedskalnin amazes us, once again, with the description of a magnetic system to create an artificial perpetual motion: the "Perpetual Motion Holder".

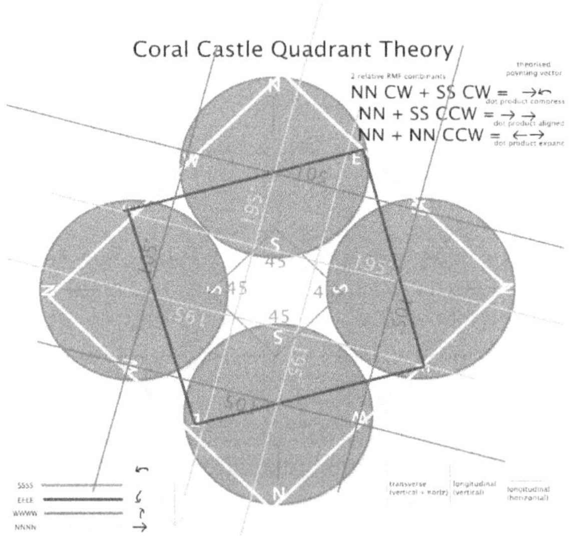

Eric Dollard, Coral Castle Quadrant Theory

CHAPTER 10

Illuminations: Sun and Moon

The Sun is certainly the star that most affects us in our lives. Every day we notice its presence or, even more, its absence during bad weather. With a diameter of 1391000 km and its enormous gravitational force, it is the fulcrum of our Solar System. The Sun is about 149 million 600 thousand km away from us and it is thought that in four billion years' time it will become a red giant with a diameter that will reach the current orbit of the Earth. Later, thanks to frightening nuclear events, it will shrink, turning into a dwarf star that will shine an almost infinite white light. The Earth orbits the Sun in an elliptical orbit: when it is farther away (aphelion) it spreads more heat to it, conversely it radiates it less when it is closer (perihelion). We have seasons thanks to the inclination of the Earth's axis of 23° 27' kept stable by the Moon orbiting the Earth at about 384 thousand km.

Since ancient times, the Moon has influenced our life with its periodic cycles. Scholars say that during the full moon period ancient men refrained from hunting, while women had their period in order to be fertile for their return. All law enforcement and E.R. workers have noticed that during the full moon, the number of operations increases considerably. Is it the hunter's atavistic instinct that affects the human soul? Surely the moon affects the oceans by causing tides. It moves away 3.8 cm every year and it is thought that in a billion years, being too far away to be subject to Earth's gravity, it will leave the Earth at the mercy of itself for good. I know what you are thinking that all this can be found in science books, but it was meant to prepare you for the alternative: the illuminations.

Yin and Yang

Let's take Gleason's map as a reference for a flat Earth and let's assume that the Sun turns clockwise on it, at the height of the Equator. On the opposite side of the Sun we set the Moon: if they rotate together it is possible to observe the effect of the day with the Sun at the centre of a halo of light and the effect of the night with the Moon at the centre of a halo of shadow. A sort of immense yin and yang, a perfect synchronized clock, where seasons are given by the Sun oscillating between the two tropics.

On its annual journey, the Sun will pass twice over the Equator and once over each of the tropics. This explains the milder climate above the Tropic of Cancer and below the Tropic of Capricorn. The phases of the Moon on a flat Earth are easily explained: the solar day lasts 24 hours, while the lunar day lasts 24 hours and 50 minutes. You only need to consult a lunar ephemeris to verify by yourself that the Moon rises every day a bit later. Being slower, it gains an average of 11 degrees of distance from the Sun every day and this causes the moon phases. The eclipse of the Sun occurs when the Sun passes directly behind the Moon. Yes, you got it right, behind the Moon, because the illuminations perform different movements with great synchronicity. An elegant and fascinating dance that repeats itself every day, month, year with the sporadic rituals of eclipses or the frequent rituals of sunrise and sunset.

Scholars of the flat Earth speculate that the illuminations work as a kind of magnetic battery, where the Sun is the positive pole and the Moon is the negative one. It has already been proven how saltwater reacts to electromagnetism: this phenomenon would cause tides in the seas without affecting freshwater lakes, which is diamagnetic. With a time-lapse photograph of the starry sky it has been noticed how the North Star remains motionless while all the other stars seem to rotate around it. Considering the incredible speeds of our Solar System in Space and the motion of the Earth around the Sun, as the months, years and centuries go by, the North Star seems to remain motionless in its place, always centred exactly above the North Pole. According to official science, the Earth rotates on itself much faster than the Moon orbits around it, but when the eclipse of the Sun occurs, the Moon's shadow moves from west to east. This phenomenon can only be explained if you think of the Sun "overtaking" the Moon by passing behind it on a stationary Earth. While N.A.S.A. still can't explain the solar eclipse of August 21, 2017, the Flat Earthers have done it in the way I have just described. Even the moon eclipse of January 31st, 2018 cannot be explained considering a globe Earth: computer simulations leave no doubt.

Four years before the supposed landing of Apollo 11, Professor R. Foster, during an ABC television interview in 1965, claimed to have evidence that the Moon was a plasma. Reporter Bob Sanders pressed him asking if it was possible to land on the moon and pointed out that both the Russians and the Americans wanted to send men to the Moon. Foster replied that if he had asked him this question about ten years earlier he wouldn't have been sure, but now, according to his scientific experiments, he was sure about it: - The Moon is a plasma and not a rock and has a mass which is a thousand times lower than the presumed one. The professor went on by saying that a good part of our science would have to be rewritten to explain tides and other phenomena. Usually, scientists do not want to commit themselves with their conclusions and they prefer using terms like "hypothesis" and "theory" when presenting their arguments; they hardly say they are sure of something. A fact that supports Foster's claims are the craters, that we can see on the Moon, because if it were true that when the Moon orbits it really shows the same portion to the Earth, the craters shouldn't appear as concentric centres perpendicular to the Earth. Impact craters leave different marks on a surface, instead the cause of these craters can only be attributed to an electrical storm. William Brian, a nuclear engineer, published a book in 1982 questioning the Moon's gravity with similar arguments.

According to both the flat Earth community and the traditional scientific community, the Moon has always remained a mystery. Many observers have reported the transparency of the Moon even when it is high during the day. The blue colour of the atmosphere in those cases seems to fill its "spots". A modern researcher has stated that the Moon is self-illuminated and that the dark parts are due to a malfunction caused by cataclysms. There have been many sightings of a planet passing in front of the Moon. J. Atkinson describes in "Earth Review Magazine" a Moon with a not perfectly opaque body, as if it were composed of a crystallized substance. This would be demonstrated by the stars that are sometimes seen shining through it. On March 7th, 1794 Samuel Bern Robotham writes in his book of four astronomers (three in Norwich and one in London) who detected and documented a visible star in the dark part of the Moon. Sir James

South, member of the Royal Observatory in Kensington, wrote in a letter to The Times (dated 7th April 1848) about the fact that he saw a star approaching the Moon and then apparently slipping inside it instead of disappearing behind it on 15th March 1848. The Moon was 7 and a half days old, but he could not determine whether the star was behind the Moon or in front of it. The same impressions were described by Thomas Gaunt in the monthly magazine of the Royal Astronomical Society of 8th June 1860.

On 24th May 1860 Gaut observed the occultation of Jupiter by the Moon. He enriched his account with technical data about the telescope that he used and described how Jupiter remained visible through the Moon, a phenomenon that lasted until Jupiter remained in the shadow.

Professor R. Foster

CHAPTER 11

Sun and Moon: let's go all back!

The exosphere, which is located after the atmosphere, is expected to expand between 2000 and 2500 kilometres, but according to a 20-year study lead by a team of Russian scientists, the atmosphere would extend up to 630,000 kilometres (Feb 2019). This would automatically incorporate the Moon into the Earth's atmosphere and would also explain why the blue sky can be seen transparently through it. - The Moon travels through the Earth's atmosphere - explains Igor Baliukin, first author on the subject, - We didn't know this until we examined the observations made over 2 decades ago by the SOHO probe.

Eric Dollard proposed to recreate Tesla's technologies in order to design a faster, lighter, leak-free power supply system. After suffering eight mysterious and unsolved fires he was forced to live in the desert. In these last four years of investigating the Sun led Eric to leave his laboratory at Sonoma State University.

His conclusions about the Sun are incredible and show that the Sun is not a fireball. According to Eric, the Sun does not have an internal structure, it is empty, it is just a surface and there is nothing inside it. Sir Arthur Stanley Eddington (Kendal, December 28th, 1882 - Cambridge, November 22nd, 1944) was an English astrophysicist who introduced the idea of nuclear fusion present within the suns and stars. Eddington's fusion was also approved by Einstein as part of his theory of relativity. To Eric Dollard the Sun was not burning anything: there was no hydrogen fusion in helium that needed oxygen to occur, and oxygen was only produced through electrolysis or photosynthesis.

Eric Dollard

Dollard said that the Sun was a transformer, a converter of energy from another dimension. He reached this conclusion because both the light and the heat produced by it were residual processes of this conversion.

The Sun is between 5000 and 6000 km from the Earth's surface, inside the ionosphere, with an estimated diameter of 5000 or 6000 km. Dollard stated that magnetic induction produced light, heat, energy, electromagnetic frequency, and radiation.

According to official science, the Sun is composed of iron, nickel, chromium, and magnesium, which are the elements responsible for its magnetic force. According to official data the temperature of the Sun is 5778 degrees Kelvin (5504 degrees Celsius). The magnetic properties of the above-mentioned elements are nullified if subjected to a temperature above 1000 degrees Celsius. Both hydrogen and helium do not produce magnetism, so the Sun cannot produce any magnetic field capable of attracting other bodies into Space. For Eric Dollard and Jose Alfonso Hernando Habecon, satellite telecommunications expert, the Sun would be cold. By generating very high electro-magnetic frequencies, the Sun causes molecules and particles of other substances in the atmosphere to vibrate. This vibration radiates heat to the Earth as it descends from above to its surface. The operation is like that of a microwave oven.

Donald Scott, PhD in electrical engineering, along with Michael Clarage, PhD in physics and electrical engineering, led the "Safire" project. They both claim that the Sun is not a ball of molten hydrogen in space and that heat is not released from it. From their research the heat would be generated by electromagnetic waves that hit the mesosphere, stratosphere, and troposphere. The higher we rise in altitude, the more the temperature of the air drops. At an altitude of 10,000 meters the temperature drops dramatically, while instead an airplane should be melted-by the Sun. According to official science, "this happens because the further we move away from the ground, the more heat dispersed from the earth is dissipated into the atmosphere and solar heating cannot heat the air effectively. The decrease in temperature and the difference between soil and high altitude is called the Thermal Gradient".

CHAPTER 12

Towards the celestial vault: new horizons

Both the Sun and the Moon keep our entire system in balance; without them there would be no life on Earth. But how are they able to they spin in our sky? Perhaps the best explanation of how this incredible planet of ours works is still provided by the electrical engineer Eric Dollard. In a few words, the Sun captures Earth's energy and that of other dimensions of space and converts them. Its buoyancy and positioning within the system are due to Earth's strong magnetic field and the electromagnetism of the celestial vault. Therefore, the Earth's magnetic field is much larger than that of the Sun and the Moon. The sky is also loaded with electric particles and filaments, which produce electromagnetic fields. The Sun collects all these electric charges: the frequencies that come from the Earth, the sky, and the celestial vault. All this energy captured by the Sun is released in the form of light, heat, frequencies, and radiation. The electromagnetic waves radiate high and low frequencies: The Sun would be the result of a vibrating frequency radiated by the Earth, which would act like a Tesla coil.

To astronomers, all the stars are light years away from our planet. The North Star, for example, would be 430 light years (2 quadrillion miles) away. To understand the celestial vault of Flat Earth, you need to think smaller, in terms of thousands of miles, or miles if you prefer. Moreover, the stars would not even be composed of gas, as we have been told. What is puzzling is the spread of the idea that the star *Vega* will take the place of our North Star in about 13,000 years. There is a lot of evidence instead that indicates that these stars are in fact a kind of timing system, if not even some kind of map given by the constellations they make up. At the centre of the celestial vault is the North Pole, which is perpendicular to the North Pole, while further down, at the height of about 3000 miles, the Sun and the Moon rotate around this axis. The two stars are the same size they appear to us and not because the Sun is immense and further away than the Moon. The rest is formed by the stars and finally, resting on the outer edge, is the South Pole. This is roughly what the magnetic motor of our house should be like. Flat Earthers admit they do not claim to know more and accuse N.A.S.A. for helping to falsify known science.

On 27th May 1931, the Swiss physicist Auguste Piccard climbed to an altitude of over 16,000 metres in a pressurised capsule designed by him. Piccard and his assistant were the first men to go into the stratosphere. They risked their lives due to a malfunction in manoeuvrability and a loss of cockpit pressure, but they managed to survive by taking measurements and bringing precious samples of ozone-rich blue air to earth.

Auguste Piccard

They had to wait for the sun to set in order to return to earth; unfortunately, they had to spend many hours in the stratosphere. They had taken off from Auguste Piccard Augsburg, today's Augsburg, in Germany, but then landed on the Gurgl Glacier in Austria. You can verify this for yourself, but because of the Coriolis effect this would not be possible at all. Due to the rotation of the Earth, about 1700 km per hour counterclockwise, they should have landed further west of Augsburg, in France or Spain, not further south than 200 km. In the first interview after their adventure, Piccard described the Earth the same way it was stated in the article of the "Popular Science Magazine" of August 1931: "It seemed a flat disk with upturned edge". If you look at the site houseofswitzerland.org they conclude a small paragraph dedicated to Piccard with these words: - He is the first man to have seen the curvature of the Earth with his own eyes – It's curious that they don't say that, according to him, the curvature was absent.

Having left from a base in the desert of New Mexico, near Roswell, Felix Baumgartner made a record flight from 39000 meters on October 14th, 2012 touching 1357.64 km/h (843.6 mph or Mach 1.24). The camera footage from the cameras outside the launch cabin had "fisheye" lenses, lenses that rounded images, providing the Earth a spherical shape. The camera mounted inside was instead equipped with normal lenses and, in the rare moments when Felix moved, you could see behind him a flat horizon. Considering the time needed by the enterprise, about 90 minutes to ascend and 4.25 minutes to descend, because of Coriolis effect, it should have landed in the middle of the Pacific Ocean and not in the Roswell desert, as it claimed to have.

CHAPTER 13
The Antarctic

Throughout history many men, supported by the governments of their nations, have tried to reach the base of the magnetic vault in the Antarctic. U.S. Navy Rear Admiral Richard Evelin Byrd (33rd degree Freemason) was the first man to fly over the South Pole on November 29th, 1929. During an interview he stated that there were immense lands to discover, larger than America. The Nazis proposed that Byrd would join their 1938 expedition, but he refused.

Antarctic is an immense ice zone, a continent with high walls perpendicular to the sea, where temperatures reach -70 degrees Celsius and life is impossible. It is assumed that on a globe Earth the temperatures are similar at the same opposite latitudes, instead at the North Pole we can find fauna and vegetation in summer and a minimum temperature of -20 degrees in winter. And yet, despite the harsh climate, many scientific and military bases are stationed in the Antarctic.

The most important nations in the world have signed the Antarctic Treaty, nations that had been adversaries in previous wars have now made an agreement; countries like the United States and Russia, which both faced the Cold War. It is practically as if they had "locked down" the Antarctic with that treaty, but why if there is nothing there?

Many politicians often go to the Antarctic, like President Obama for example, but the real reasons for these visits remain secret. If you tried to apply for permission from your own state, you would find yourself in a very long bureaucratic process, which would almost certainly end with a negative outcome. You are only allowed to access the country if you are a scientist, or if you join some expensive guided tours. But let's suppose you're lucky enough and you get the permission to enter the South Pole autonomously: no motor or sailing vehicle can cross the 60th parallel South, not even a raft or a rowing boat; you cannot even swim.

But you are lucky and still manage to arrive on Antarctic soil; for your expedition you cannot use motorized vehicles because only the military are authorized to lead them. You can also forget about an animal-drawn sleigh; in truth, you are not even allowed

to use a sled pulled by yourself. As a matter of fact, all you must do is grabbing supplies enough for a month and walking on inhospitable ground beyond all human limits. Don't forget to bring your own poop storage bags, as it is forbidden to get the soil dirty.

Highjump

The Americans organized in 1946 the operation "Highjump" commanded by Richard Cruzen, organized and co-directed by Byrd, which included: the ship Mount Olympus, the aircraft carrier Philippine Sea, 13 support ships with dozens of planes, seaplanes and helicopters. About 4500 men took part in it, divided into three groups: eastern, central and western.

Today "Highjump" is the most impressive Antarctic expedition ever made. They arrived in the Ross Sea on December 30th, 1946. On 11th February 1947 they discovered the Bunger Oasis, named after the aviator David Eli Bunger. He declared in his report that he had found a huge area of land with freshwater lakes in the ice. Later they returned to the oasis and landed on one of the lakes. The water measured 30 degrees and contained red, blue and green algae. The most important discoveries of the operation "Highjump" were omitted. Byrd's family also complained about the fact that one of his precious diaries disappeared. The admiral's son described him as a meticulous and precise man, very tidy. Fortunately for us, the forward-thinking Byrd had transcribed part of it in the blank pages of the diary of his previous mission in 1926. Thanks to this stratagem, we learned of his Antarctic flight on February 19th, 1947, where Byrd, along with his assistant in charge for communications, found other green lands, with animals that were thought to be extinct. His plane was intercepted and, without him piloting it, landed in an underground city which, even in the absence of Sun, was somehow illuminated. Byrd describes it as a magnificent city. They were both welcomed, but only the Admiral was accompanied to a powerful man. In English, but with a Nordic accent, perhaps German, the Admiral were assigned the task of bringing a pacifist message to his government. In practice, the earthlings were asked to stop using nuclear power for war purposes, or his technologically advanced and fearsome people would intervene. Obviously, after reporting this to his superiors, Admiral Byrd was ordered to maintain the strictest secrecy on the matter.

Personally, I cannot call a man, who had such a great reputation, that he was assigned to such an important mission like Highjump, delusional. I saw Byrd's television interview and you can immediately understand what kind of man he was; you can see it in his eyes: determined, precise, reliable, calm but ready for action, with the spirit of the explorer in his blood.

Operation Highjump, which had to last from six to eight months, was interrupted after only eight weeks for unknown reasons. In the years that followed, the U.S.A. organized other expeditions including the most important one: Operation "Deep Freeze" in 1956, a collaborative effort between forty nations. Not surprisingly, the following year the first space satellite, Sputnik, was launched into orbit, probably because they had discovered the limit of Earth's electromagnetic field and wanted to probe it. In 1958, again not by chance, N.A.S.A. was founded to give us the illusion that we could get out of Earth; N.A.S.A. was trying to hide from all of us the fact that this vault existed. Stipulated in Washington on December 1st, 1959 and signed by 53 countries participating in the International Geophysical Year (1957-58), the Antarctic Treaty came into force on June 23rd, 1961. "Art. VI:

the area covered by the treaty extends south of the parallel with latitude 60° S. It includes the entire ice cap".

In my opinion, this pact of non-belligerence, respected by everyone below the 60th parallel, is the consequence of the pacifist warning made by Byrd.

Practically all the states in question, since then, have continued to declare war and to do their policy of convenience, but only above the 60th parallel South, far from indiscreet and fearsome eyes. Of course, everyone, in principle, agreed to proceed with nuclear disarmament.

Richard Evelin Byrd

CHAPTER 14

N.A.S.A.: retouched photo

As a child I was proud of my globe: a perfect sphere with a magnificent glass and a light bulb that illuminated it from the inside. I knew that the shape was not the right one because my science teacher had taught me that the Earth is crushed at the poles. The South Pole is slightly more flattened than the North Pole by a difference of 15 meters between the two polar rays. This crushing of the Earth is due to centrifugal force, she told me. Today I know for sure that, rotating at about 1700 km per hour, all the water of the oceans would converge at the Equator and then disperse into Space. Supposing that the force of gravity, in its role, really holds back the water of the oceans and that the Earth is crushed at the poles, at the Equator you would have the apex of a convex line higher than Mount Everest. The River Nile, which crosses the Equator, would miraculously flow uphill at that point.

- But we've been to Space, we've got pictures! - I used to repeat to myself, - There is the International Space Station (ISS) that has been orbiting the Earth for years, we have proof! -. And yet, even from the recent Chinese, Indian and Israeli space missions, certain evidence has emerged, thanks to the analysis of these skeptics, that reveals photomontages and numerous inconsistencies in the videos.

We have a lot of technology at our disposal and we should just use it to verify the authenticity of a photo. For example, it is just enough to increase the saturation of the picture: if square contours appear around a spherical object, it means that the image has been superimposed. Both the Moon and the Earth appear to have been added later in pictures that were supposed to have been taken in Space. In the first photo of our planet, the Earth appears to be a perfect circle, which is unreal since it is a snapshot. In subsequent photographs certain continents seem disproportionate, some much larger than others. N.A.S.A. claims to have hundreds of instruments orbiting our planet every 100 minutes and sending data to Earth that are processed into images. In an interview granted by a N.A.S.A. employee Robert Simmon explained: - First we take the data and extract the images; we must remove the clouds -. Then he described the choice of colours for the water of the oceans that are derived from instruments for measuring phytoplankton in the seas. - Where the level is low, I add some blue, because it is usually blue, then the bright green -, this colour refers to the vegetation on land.

Then the clouds are added, this is usually problematic because there's a slit in the eye sockets. Some of these are painted... Eh! Yes, it's Photoshop, but it must be! -You got it right, it's Photoshop, but it must be! That's our Bob! He goes on listing the levels he uses to simulate the atmosphere: I avoid this martyrdom. The community of the Flat Earth certainly did not need Simmon's explanation to be sure that the N.A.S.A. photos are not real. From Hanlon's razor: - Never attribute to bad faith what can reasonably be explained by stupidity -. In our case I see bad faith applied with stupidity and with arrogance, given the low opinion N.A.S.A. has of all of us.

The blue marble

Taken on December 7th, 1972 by the crew of Apollo 17, at about 45,000 km, the Blue Marble is perhaps the most beautiful photograph of the Earth that N.A.S.A. has provided us. According to N.A.S.A., the official photograph of the Earth is the one depicting the two Americas. Now I'm going to prove to you that this photograph is a fake achieved through a video, you can verify this for yourself.

Let's take as an example the picture where Central America stands out, then we will take as reference two well visible locations in Mexico: Puerto San Carlos and Tamaulipas. The first overlooks the Pacific Ocean and the second one the North Atlantic Ocean. According to Google Earth, from coast to coast, the two locations are divided by and large by 1503 km. The official diameter of the Earth is 12742 km. We now have enough to expose the scam of N.A.S.A. once and for all. Locate the two points in your picture of the Earth and cut out a corresponding piece of paper. In order to cover the entire diameter, you must cover the Equator eight and a half times. Place the strip of paper at the height of the equator and you will see with your own eyes that the diameter of your Blue Marble is only six times the diameter of your Blue Marble. Like a street hustler, N.A.S.A. has dumped a fake marble on all of us and sent only our imagination into Space. There have been many protests over the Space Agency because of this. In order to mitigate an emerging discontent with them, N.A.S.A. turned the MRO satellite in

Martian orbit towards Earth. From 200 million kilometres the MRO took a photograph of our home. Of all the pixels that make up the picture, do you know how many of them Earth is made up of? One, just one pixel.

I challenge any of you to recognize a person's face in a photograph where his or her face matches to a single pixel. But according to N.A.S.A., even if electronic intervention is allowed, it is possible to distinguish in that pixel some portions of terrestrial continents. Here is the official version: "The photo is the result of two shots with different exposures. In fact, the Earth is much brighter than the Moon. This would therefore be too dark. The image has been reassembled respecting the size and position of the planet and its satellite. From Mars you can see Australia, the Indochinese peninsula and the white glow of Antarctic. In this beautiful postcard sent from over 200 million kilometres Mars Reconnaissance Orbiter shows us its point of view of our planet and the Moon, photographed together on 20th November 2016".

1500 KM
Diego
Phoenix
Dallas
MISSISSIPPI
Tucson
El Paso
TEXAS
BASSA
CALIFORNIA
Austin
LOUISIANA
SONORA
Houston
New Orleans
San Antonio
COAHUILA
1,000.00 km
500.00 km
1500 km
Monterrey
SINALOA
DURANGO
TAMAULIPAS
Messico

CHAPTER 15

Video & Freemasons

"Edwin Eugene Aldrin, Jr., aka 'Buzz', member of Clear Lake Masonic lodge no. 1417 in Seabrook, Texas. Friar Aldrin brought with him a special proxy by which the Grand Master J. Guy Smith appointed him as special delegate of the Grand Master himself, granting him full power to represent him on the site and authorizing him to claim Masonic territorial jurisdiction over the Moon on behalf of the Most Venerable Grand Lodge of Texas, Ancient, Free and Accepted Masons" (from the archives of Tranquillity Lodge 2000 of the Grand Lodge of Texas, A.L. & A.M.).

In the first mission of the alleged moon landing, astronauts Buzz Aldrin and Neil Armstrong were to mount the American flag together. From the film it seems so, but in a porthole reflection you can clearly see the flag already planted on the lunar ground before Buzz reaches his companion outside the LEM. Buzz, as Freemason, was in charge of performing a symbolic rite when he erected the flag on the lunar ground.

Whether they were on the Moon or not is irrelevant: he had an obligation to do it. Freemasonry, in general, is a confraternity with rules that aims to have acolytes in key roles in the world. When there was the opportunity to send a Freemason to the Moon, the leaders of that time acted with a delegation to Buzz Aldrin. Being a Freemason is not a bad thing and even ordinary people like us can join, starting at the bottom rung. The levels range from the first degree to the thirty-third. For people who are not initiated, Freemasons have an obligation of secrecy, but also an obligation of truth. They must therefore out of necessity leave us little clues, so as not to contravene this second obligation. In the logo of N.A.S.A., for example, one sees a red "V" lying on the left: this is a Masonic symbol that indicates the presence of brothers among the staff. The paintings depicting famous people, an open hand resting on the belly, fingers stretched out with some of them joined together, reveal the belonging to Freemasonry. If these two men have never set foot on the Moon, one thing is certain: Freemasonry, like us, has fallen victim to this deception.

In 1969 N.A.S.A. had used the best technology at its disposal. Fifty years since then, each of us, in our own homes, has far superior means and instruments at our disposal. Just a bit of specific knowledge is sufficient to analyse the material in its possession. But N.A.S.A., the repository of man's greatest enterprise, battens down the hatches and tells us that we can no longer find the Apollo 11 mission tapes. They are practically there, they are still there, but they are not there anymore. You see how the lie repeats itself over time? Just like dark matter, which is there but cannot be seen, the tapes are there but cannot be found. The icing on the cake is certainly not missing because even the technology used to go to the Moon has been lost. Luckily, China arrived on the Moon with the Chang'e-4 spacecraft and the Yutu-2 rover on January 3rd, 2019. It can be deduced that the Chinese have in their possession the technology necessary to face space travel. Too bad they went down on the "hidden face" of the Moon, completely invisible to terrestrial telescopes. Congratulations to the Chinese: they are fast workers!

CHAPTER 16
ISS: risking everything

The thermosphere ranges from 90 km to about 600 km from the Earth's surface with a temperature ranging from 200 to 2000 degrees Celsius. The temperature varies depending on the height and activity of the Sun. The International Space Station is maintained in an orbit approximately 400 km from Earth. The temperature at this height is 1500 degrees C (1221 degrees F). Aluminium melts at 660 degrees C, gold melts at 1064 degrees C, steel melts above 1370 degrees C. All electronic devices suffer from high temperatures. How N.A.S.A. can protect the ISS from these extreme temperatures seems a miracle of material science. One wonders what kind of protection the Apollo spacecraft has been equipped with in the past to pass through the thermosphere unharmed. Between 1,500 and 40,000 km the Van Halen range is included, and that is radioactive and lethal to humans. According to N.A.S.A., the astronauts have escaped unscathed through it twice, on their way there and back.

The Shuttle Challenger, which really did reach Earth's low orbit, has been hit several times by paint particles that were detached from other ships. In 1983 it became necessary to replace some of the ship's glass, damaged by small debris. The solar satellite Max, retired in 1984 after 4 years in Space, reported an average of 37 holes per square meter, all due to space debris. According to the United States surveillance network, more than 21,000 objects over 10 centimetres wide orbiting the Earth have been tracked. Smaller objects are too small to be detected. It is estimated that there are another 500,000 pieces of debris ranging from 1 to 10 centimetres of size. In addition to these, particles smaller than one centimetre would be several million. N.A.S.A. has stated that an impact in space is equivalent to a flat impact on the ground. An object only 5 centimetres in size has the same energy as a 16.5-ton bus. A 10-centimeter object has the same kinetic energy as 300 kilograms of TNT. The ISS has a speed of 28,000 km/h and in a head-on collision the impact speed has to be doubled. The ISS orbits the Earth in 90 minutes, while an average detritus takes 70 minutes. The duration of these relative short orbits significantly increases the possibility of an impact. Yet the ISS has been in orbit for years and has never suffered the slightest collision, not even one prevented. From the window of the ISS, you can see clouds, land and sea with a wealth of detail, yet you never see a satellite, an aircraft flying or space debris passing through.

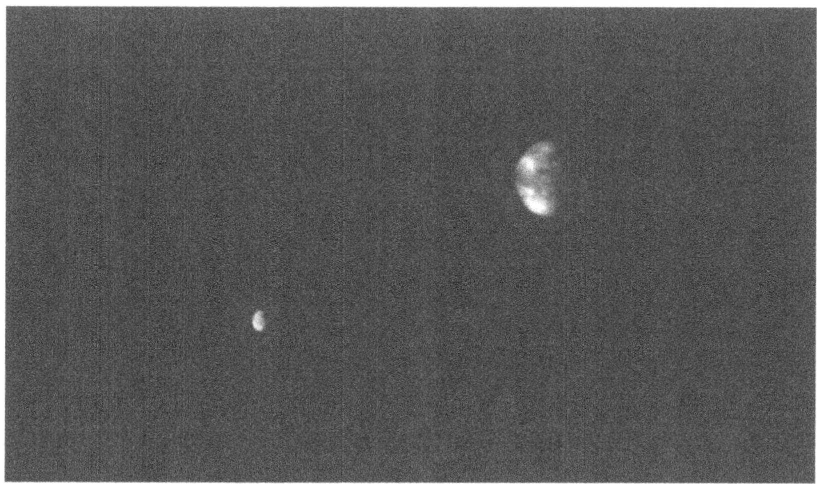

Photo by MRO (Chapter 14)

CHAPTER 17

A dive in the pool

During training, astronauts train in a pool to simulate conditions in Space. The pool is located in Texas, in the Neutral Buoyancy Laboratory of the Sonny Carter Training Facility, located at the Johnson Space Centre in Houston. The NBL is the largest pool in the world, where the entire ISS habitat is developed. Some scuba divers assist astronauts during operations. It's a shame that a lot of the footage made during training has been electronically modified to make us believe that everything happens in Space. You can see in those videos strange air bubbles floating in space. You can even see a diver, wearing a mask and tanks, reflected in an astronaut's visor. On live TV recordings, some astronauts of the ISS seem to handle objects: they often twirl them in front of the camera in the apparent absence of gravity; too bad that these objects sometimes disappear or get confused with the environment, revealing the trick. The astronauts' long hair is clearly full of hairspray and "shot" at the top, in order to simulate its fluctuation.

Behind the astronauts the Earth's rotation seems to stop or even to go backwards. Sometimes the Earth increases its rotation speed, while the clouds do not follow it or stop altogether. In one video, behind the astronauts, you can see an open skylight, while in another one an astronaut becomes transparent, blending in with the walls of the ISS. In some cases, wires have been seen from which the astro-no-nauts would hang. Every now and then there is a strange interruption of weightlessness, perhaps due to the fact that they are at that moment inside a zero G plane. For those who don't know what that is, zero G planes are aircrafts that create absence of gravity by going into free fall.

The U.S. Senate has awarded N.A.S.A. $22.6 billion in funding for the calendar year 2020. Bearing in mind that to make a good science fiction film like "Gravity" it required a budget of $100 million, don't you think N.A.S.A. could do better than that?

CHAPTER 18

Shuttle: "mission impossible"

In the video of the Shuttle throw in 1983 you can see the astronaut practically without a spacesuit: between the helmet and the suit you can see the bare neck, while the gloves are not hooked to the sleeves, revealing the forearms. But without the pressure suit, isn't the astronaut putting his life at risk?

On January 28th, 1986 the Shuttle Challenger exploded 74 seconds after its launch, killing the 7 astronauts on board. All nations expressed their condolences to America and paid tribute to the lamented astronauts around the world with a day of national mourning. SURPRISE: Six of these dead astronauts were found alive! All of them denied the evidence, of course! From 1986 to the present day their appearance has changed, but their facial features still stand out. Many believe this news is not reliable, but I find it suspicious, more than the possible similarities, that the presumed astronauts have kept the same names or surnames.

The driver of the Challenger, Michael John Smith, has now become Professor Michael J. Smith of the College of Engineering, University of Wisconsin-Madison. Cargo specialist Christa McAuliffe now uses her middle name Sharon and works as a professor at Syracuse University.

Ellison Onizuka, mission specialist, has become his twin brother Claude.

Mission Commander Francis "Dick" Scobee is now just Dick Scobee, CEO of a real estate agency.

Mission specialist Judith Resnik, after a skin whitening treatment, did not even bother to change her name and she is now a professor at Yale Law School.

Another mission specialist, Ronald McNair, has become his twin brother Carl S. McNair.

Let's remember that Shuttles, unlike the International Space Station, have carried out real missions in Space facing all the risks this entails. The prerogative of the Shuttle was to be retrieved after each mission by reducing its costs. Nevertheless, the missions remained highly expensive and the discontent of the American taxpayers in the 1980s was palpable. Probably the leaders of N.A.S.A. considered different solutions before choosing the one we all know. Shuttle missions were discontinued for two years, benefiting the space agency's budget.

I think they convinced the Challenger's crew with a national security speech filled with patriotism. It was an explosive ending, speaking in terms of fireworks.

CHAPTER 19

Stars' "ON/OFF"

It is believed that from Space you have a privileged vision to observe the stars, so weird that during the Apollo missions no photo was taken to immortalize them. In the first interview to the Apollo 11 astronauts, after 30 days of quarantine since their return to Earth, you can clearly observe that they did not want to be there, in front of those journalists. All three astronauts look strangely disconsolate and shy, while one would expect to see some enthusiasm and healthy pride in them. When the journalists asked them what the stars looked like, Armstrong replied uncomfortably, saying he had not seen any. Aldrin tried to help him, by saying he had not seen any stars either. Collins also confirmed the two colleagues' version. Now "Houston, we have a problem!": the navigation system of Apollo 11 was oriented to the stars and was equipped, for this purpose, with a space sextant called "space sextant". Buzz Aldrin himself showed during a television interview the very star map used to orient himself in space during their mission.

However, many other astronauts, such as Edgar Mitchel of Apollo 14, have reported that they have not seen stars in cislunar space, but only deep black. Here too, we have a white fly: astronaut Chris Hadfield, live on TV from the ISS, described the panorama:- The sky is almost completely white due to the lights of the Universe, with an incalculable number of stars...you can see the constellations because the sky is so alive! -. Another astronaut, Tim Peake, always in person from Space: - An unexpected thing is that Space is completely dark – But. What? first Space was dark, then completely white and now dark again? Astronaut James Reilly talked again about stars in a television interview: - ...You can see stars and colours that you can't see here on the ground, pastel colours, one yellow, others orange and even red and dark blue... Colours... -. Although his astronaut colleague gave him comfort, host of another TV show, Chris Hadfield changed version: - ...And what you see is...(pause) It's an immense deep darkness. The darkness is like a black, almost purple texture where practically nothing is reflected -. Still a guest on another TV show: -...When the world.... Up there... You look the other way, you see the whole black, black universe, palpable, simply dark, and... Forever and ever –

CHAPTER 20
Vacuum-sealed

At home many of us are familiar with vacuum food storage. In this case, the sucking of air is thought to be negative for when it concerns the Earth's atmosphere. Instead, it is assumed to be positive if we compress air to inflate a ball or tire. Speaking of a vacuum, the lower the number the more you are subject to negative pressure. According to N.A.S.A. the minimum we find in outer space is 10 to -6 Torr up to a maximum of 10 high -17 Torr (an atmosphere corresponds to 760 Torr). These are impressive figures, considering that the tyres of an average car need a positive pressure of 2.5 atmospheres (1900 Torr). Before giving you more comparable figures, let's see how the world's most famous space agency tested the vacuum on its equipment.

The Space Power Facility is the chamber built by N.A.S.A. to recreate the vacuum. The construction was completed in October 1969, three months after the astronauts had already gone to the Moon.

This shows a certain rush to take off at all costs and, and if considering the fact that none of the three astronauts ever swore an oath on the Bible to authenticate the mission, it all looks very suspicious. Can we believe that the space engineers planned the first trip to the Moon without first experiencing the conditions in the void? Without first testing the spacecraft, the LEM and the spacesuits? The SPF is 37 meters high with a diameter of 30 meters and can withstand up to 10 to the -6 Torr, which is the minimum for the vacuum in Space. The reinforced concrete walls range from 1.8 to 2.4 metres thick. Air, with a vacuum of 10 to -6 Torr, could still be sucked through the walls, so the structure would be reinforced with a watertight barrier formed of steel drowned in the containment barrier.

Men have walked on the Moon, so how did they counteract the vacuum of space outside the LEM? On the N.A.S.A. site there is uploaded a table with all the materials that make up an extravehicular suit resistant to vacuum and solar radiation. Let's see this list together: rear entrance zipper, Teflon fabric, beta fabric (against abrasions and flames), Kapton aluminized plate (reflective insulation), Kapton laminated siliceous fiber fabric (located between the reflective layers), Mylar aluminized (5 layers of reflective insulation), non-braided fabric Dacron (spacer), neoprene nylon fabric, nylon (to withstand pressure), etc...

You can consult the rest for yourself, but you will not find thick concrete barriers and an airtight chamber to complete the suit. For N.A.S.A. this suit, with a pressure of only one atmosphere inside it, is enough to counteract the vacuum on the Moon which is 10 high at -11 Torr, one hundred thousand times stronger than the maximum obtained on Earth by the SPF. Think of a single atmosphere that contrasts 136 million negative atmospheres (almost 2 billion negative psi). The seals used in Apollo missions, both for helmets and gloves, are inadequate even for low pressure. Metal gaskets, specific for high pressure, would be needed to ensure optimal sealing, and surfaces should be perfectly clean. The same applies to the lunar module hatch because the astronauts would find themselves in the dust, dragging dirt, and the lunar module door seals would not be able to function properly. In addition, the LEM was without a decompression chamber, which means that once the hatch was opened any container inside it would explode food, water, bags of urine, feces, and air tanks. A $10,000 prize was offered for anyone who wanted to wear a suit inside the Space Power Facility. Until today, no daredevil has honoured this challenge.

During an interview Terry Virts, a former N.A.S.A. astronaut, said: - N.A.S.A. is planning to build a rocket called the SLS, a heavy-duty rocket that is much bigger than the ones we have today. [...]It will be able to launch the Orion capsule with men on board as well as vehicles to land [on planets], and other instruments to destinations beyond the Earth orbit. Currently we are only able to fly within the Earth's orbit, further than this we cannot go -.

N.A.S.A. is compartmentalized and, as it happens in any large company, only an executive can have the full picture of the operations that take place within it. Those who work there, carry out their work honestly, unaware of the decisions taken at the top. Among the many perplexities expressed towards N.A.S.A. is the absence of a video recording an astronaut using the decompression chamber before a spacewalk. It is fair to believe that for an astronaut this is the most exciting moment, where he can make a good use of his training. You only see astronauts using it in science fiction movies, how is that possible?

According to Sylvain Timsit, mentioned earlier in the first chapter, one way to control the masses would be to postpone the solution of a problem. Basically N.A.S.A. tells us that today it is not possible to go to Mars, but tomorrow it will be possible. So, with this simple promise, the executives of the space agency will be grabbing billions of dollars in subsidies until the expected year for the Martian mission: 2033.

CHAPTER 21

Everyone connected, why?

When you reach the last chapter of a book, readers are likely to expect a final conclusion, but in this particular case the epilogue concerns our lives and we are the only authors.

In 1981 Commodore introduced the VIC-20, its first machine to be defined a home computer, equipped with a 32K memory card with the possibility to add an expansion of 3K, 8K, 16K, 32K, 64K. Then, in 1982, the Commodore 64, which made history, made its entrance. In 1984 the Commodore 16 was released and in 1985 the Commodore 128. With Amiga line ended the story of the Commodore, which was mainly replaced by Apple Macintosh. What do these numbers tell us? Let's take the SIM card of our iPhone or smartphone: we started with 8K, then replaced it by 16K, 32K, 64K, 128K SIM to arrive at today's 256K. The same thing happened to SD cards, Micro SD cards, USB sticks, whose numbers are expressed in Gigabytes instead of Kilobytes: 2GB, 4GB, 8, 16, 32, 64, 128 and today's 256GB.

They can be RAM, ROM memories, graphics or sound cards; we always have the same modus operandi even in PCs. I'm not saying that they've been feeding us with the same technology for forty years, but the doubt tortures me. Surely, they are selling us new products using the same numbers. It is as if a climber from the past had mapped out a route and the modern climber, while possessing state-of-the-art equipment, continues to follow it without daring more. Anyway, for forty years they have been selling us at a high price memory inside pieces of plastic with a production cost of a few cents. Of course, we all pay for the research and for the processing behind, but the real added value of an object is given by our desire to own it. For the manufacturers of technological trinkets, it is only important that the production is logistically possible and economically viable. At this point it is necessary to pay the right amount of attention because one freedom often implies the absence of another: let's think about who really benefits from the new technology that is offered to us. Through all the "social" we have the illusion of feeling free and connected to the community, but in reality, we are bound to a massive control system that operates through us. In practice we control each other, day after day, and we also feel happy to do so.

You may be wondering where I am getting at with this simple reasoning: I hope I have shown you that in order to understand the delicate mechanisms of our unstoppable modern society, we often need to take a look at the past in order to be guided towards the future.

Foreboding

On the basis of what was said before, we should keep an eye on the Deagel site where 209 nations of the world are listed, along with all the data relating to their armies, economy, population, ships and more. The managers of the site take the information from banks, civil administrations, all government organizations and, after an analysis, provide their projection for the future. This data refers to 2017 with forecasts for the year 2025. For example, the population of the U.S. will increase from 327 million today to 100 million in 2025. Where will 227 million Americans go? Italy will go from 62 million today to 44 million, 18 million less Italians! In Russia the population remains unchanged, as well as in China. In Canada, on the other hand, they expect a decrease from 36 to 26 million inhabitants. In France from 67 to 39 million. In Germany from 81 to 28 million.

But how is it possible to obtain such forecasts? They are probably aware of information, data, that has not been disclosed.

I have tried to hazard a few conjectures but none of them appear to be exhaustive or possible. A world-class cataclysm, for example, is to be excluded because it would not hit so selectively. Unless someone found a way to control the climate, it would be impossible for anyone to dispense targeted cloudbursts. Wars can only be localized, and they can only affect the concerned countries, but it would only represent the solution if it did not result in the deaths of millions of people in nations so far away. Moreover, I doubt that the United States would suffer such huge losses in this case. I feel optimistic and choose to discard the arrival of Armageddon, a Hebrew term that indicates, in its broadest sense, the Apocalypse itself. It could instead be a selective and programmed exodus of populations to another planet or another vault close to ours. But something still pushes me to direct my thoughts towards the minds of the past, especially those who shared the idea that an underground people could actually exist. By exploiting the "Hollow Earth Theory ", the underground world discovered by Admiral Byrd, perhaps the legendary Agarthi, will be revealed to the people.

CREDITS

AIF Associazione per l'Insegnamento della Fisica - Oxford Academic (website) - Dino Tinelli (Youtube channel) - Terrapiatta Globo vs Terrapiatta (website) - Manly P. Hall - Cosmologia Zetetica - 200 tests by Eric Dubay - Mr. Trhive And Survive - Wikipedia l'Enciclopedia libera - Focus, Focus jr. - Avalon Island (site) - N.A.S.A. (official site) - Gak (site) - P.L. Somma - Flat Earth Brothers (site) - Robert Simmon, N.A.S.A. employee - Eric Dollard - Tesla Seismic Warning System 2.0 - The Future Past of Electricity- Mondi oltre i poli, F. Amadeo Giannini - Feed Your Mind - Flat Earth Tribe (site) - Moongate: (book) by William L. Brian - Isola Di Avalon (website) - Google Maps - Global Grey ebooks Curious Life (Youtube channel) - Margherita Campaniolo (website) - Our way is the highway (website) - Sylvain Timsit - Paul on the plane (website) - Il Giardino Degli Illuminati (website) - Dynamic Earth, book by Montgomery Childs and Michael Clarage - Luogocomune (website) - StolenHistory. ORG (website) - The Library of

Congress Mental Hygienist (Youtube channel) - Mark Knight - ODD Tv - Nextme (website) - J. Atkinson - Universe Astronomy, Barbara Bubbi - Pinterest (website) - Bitchute (website) - Norman B. Leventhal Map at the BPL.

THANKS

Sincere thanks to Professor Nicoletta Bertorelli, for her valuable contribution to the analysis of my work. I thank my children and friends who have supported me throughout the development of the book. A special thanks goes to the competent staff of Youcanprint.

ADDRESS

MAIL: <u>lucabertorelli@virgilio.it</u>

INDEX

Youcanprint

Finished to be printed in July 2020